AF228840

These are turtle eggs.

5

This is a baby turtle.

Baby turtles walk to the sea.

Baby turtles swim.

11

Baby turtles grow big.

13

Turtles live a long time.

15

Turtles eat plants.

17

Turtles sit on the
beach.

19

Turtles have shells.

21

Sea turtles live
in the sea!

23

Published in 2026 by The Rosen Publishing Group, Inc.
2544 Clinton Street, Buffalo, NY 14224

First Edition

Editor: Theresa Emminizer
Book Design: Jeffrey Taylor

Photo Credits: Cover SeanScottPhotography/Shutterstock.com; p. 3 IainCampbellImages/Shutterstock.com; p. 5 Kalaeva/Shutterstock.com; p. 7 Martin Pelanek/Shutterstock.com; p. 9 Agus_Tri1975/Shutterstock.com; p. 11 Christopher Teoh/Shutterstock.com; p. 13 Andrea Izzotti/Shutterstock.com; p. 15 MPH Photos/Shutterstock.com; p. 17 Andriy Nekrasov/Shutterstock.com; p. 19 Melissa Burovac/Shutterstock.com; p. 21 SaltedLife/Shutterstock.com; p. 23 David Carbo/Shutterstock.com.

Cataloging-in-Publication Data

Names: Emminizer, Theresa.
Title: Green sea turtles / Theresa Emminizer.
Description: Buffalo, New York : PowerKids Press, 2026. | Series: Superstars of the sea Identifiers: ISBN 9781499451436 (pbk.) | ISBN 9781499451443 (library bound) | ISBN 9781499451450 (ebook)
Subjects: LCSH: Green turtle--Juvenile literature.
Classification: LCC QL666.C536 E46 2026 | DDC 597.92'8--dc23

Manufactured in the United States of America

Some of the images in this book illustrate individuals who are models. The depictions do not imply actual situations or events.

CPSIA Compliance Information: Batch #CSPK26. For further information contact Rosen Publishing at 1-800-237-9932.

Find us on 

PK Beginners

Titles in This Series

ANGELFISH
GREEN SEA TURTLES
JELLYFISH
NARWHALS
SEA STARS
SEAHORSES

PowerKiDS press™

ISBN: 9781499451436